Understanding the Elements of the Periodic Table™

FLUORINE

Heather Hasan

9   19

F

rosen publishing's
rosen central

New York

To my son Matthew James Hasan. You have such a sweet spirit.

Published in 2007 by The Rosen Publishing Group, Inc.
29 East 21st Street, New York, NY 10010

Library of Congress Cataloging-in-Publication Data

Hasan, Heather.
Fluorine / Heather Elizabeth Hasan. — 1st ed.
p. cm. — (Understanding the elements of the periodic table)
Includes bibliographical references and index.
ISBN-13: 978-1-4042-1005-9
ISBN-10: 1-4042-1005-9 (lib. bdg.)
1. Fluorine. 2. Chemical elements. 3. Periodic law. I. Title. II. Series.
QD181.F1H358 2007
546'.731—dc22

2006009938

Manufactured in the United States of America

On the cover: Fluorine's square on the periodic table of elements. Inset: The atomic structure of fluorine.

Contents

Many of the world's greatest discoveries have been made by accident. Serendipity, or accidental discovery, actually plays a big part in scientific research. One of the most profitable discoveries ever made involved serendipity, fluorine (F), and a man named Roy Plunkett (1910–1994). Plunkett was working for DuPont, a chemical company founded in 1802 that uses science to enhance the quality of products. In 1938, Plunkett was working on the development of chlorofluorocarbons (CFCs), chemicals that contain the elements chlorine (Cl), fluorine, and carbon (C). Plunkett decided to see what would happen if he mixed a chemical called tetrafluoroethylene, or TFE (C_2F_4), with another chemical called hydrochloric acid (HCl). He set up his equipment so that the gaseous TFE would flow into a container holding the hydrochloric acid. However, when he opened the valve on the tank to let the TFE gas escape, nothing came out.

Plunkett could have thrown the tank away, but he sawed it open to see what had happened. Inside, he found that the gaseous TFE had turned into white powder. The TFE had polymerized, or joined together, into a single mass. He sent the mystery powder to some other scientists, who studied it carefully. They found that the curious powder had some remarkable properties. It would not dissolve in anything, it was not affected by heat, and it would not burn. Amazingly, it also had the ability to remain flexible down to −400° Fahrenheit (−240° Celsius), and very few things

Pictured here is a reenactment of Roy Plunkett's discovery of Teflon in 1938. Plunkett *(far right)* and his colleagues are shown with the mysterious white substance, which would later be called Teflon.

stuck to it. Plunkett had discovered a plastic called Teflon! DuPont registered the Teflon trademark in 1945 and released its first Teflon products a year later. Since that time, most kitchen cupboards contain skillets with cooking surfaces covered with Teflon. Food will not stick to it, so Teflon pans need no oil or butter. Teflon is also used in baking sprays and as a stain repellent for fabrics. Because of its useful properties, it is also employed to make things such as tubes and motor gaskets that must not deteriorate. We can thank Plunkett for his curiosity and for giving us just one example of the many interesting ways that we can use fluorine. Thank goodness for accidents!

Chapter One
What Is Fluorine?

Fluorine, whose chemical symbol is F, is a pale, yellow gas. It is poisonous and corrosive and has quite a sharp, disagreeable odor, similar to that of a mixture of ozone and chlorine. Fluorine is very dangerous. Breathing just a tiny bit of fluorine can damage the lungs. However, fluorine is not all bad. Fluorine is the most reactive of all the elements. This means that it combines with other elements very easily. Fluorine's reactivity makes it a very useful element. It is used to make toothpaste and nonstick surfaces for frying pans, and to refine uranium fuel that is utilized in nuclear power plants.

Where Fluorine Is Found

Fluorine is present in all living things. In small quantities, fluorine is found in bones, teeth, blood, urine, saliva, sea water, eggs, and hair. Fluorine is also found in the earth's crust. Because of its reactivity, fluorine is never found by itself in nature. It is always found combined with other elements. In combination with other elements, fluorine makes up about 0.065 percent of the earth's crust. Most of the fluorine in the earth's crust is found in the form of minerals.

In small quantities, fluorine is found in minerals called apatites. These minerals are made up of tiny amounts of fluorine combined with

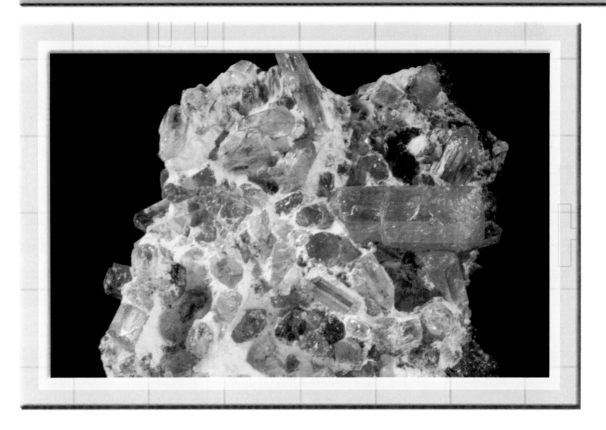

Apatites are minerals that consist of calcium, phosphorus, oxygen, and tiny amounts of fluorine. Apatites occur in igneous and metamorphic rocks, usually in small quantities.

three other elements: calcium (Ca), phosphorus (P), and oxygen (O). Apatite deposits can be found in the United States, North Africa (particularly Morocco, Algeria, and Tunisia), and Russia. A better source for fluorine is a mineral called fluorspar, or fluorite (CaF_2). This glassy crystal is made up of fluorine and calcium. Large quantities of fluorine are found in fluorspar. Fluorspar occurs in granite (igneous rock) and in limestone (sedimentary rock) and fills the cracks in sandstone. Mineral deposits of fluorspar are found in China, South Africa, Mexico, France, and Russia.

The name "fluorine" comes from the Lain word *fluere*, which means "to flow." Fluorine most likely got its name from the use of fluorspar in refining metals. Fluorspar has been used for thousands of years to remove impurities from metals, such as iron (Fe). When fluorspar is added to molten metal, impurities "flow" away from the metal and react with the fluorine in the fluorspar.

The History of Fluorine

Georg Bauer (also known as Georgius Agricola, 1494–1555), a German mineralogist, was the first to describe in detail the mineral fluorspar and its use for purifying metals. Heinrich Schwanhard, a German glass worker, discovered a second use for the mineral in 1670. He found that glass can be etched when it is exposed to fluorspar that has been treated with an acid, such as sulfuric acid or nitric acid. By 1771, scientists had learned how to make an acidic gas called hydrogen fluoride (HF) by heating fluorspar with a strong acid. Many researchers experimented with this very toxic gas. They knew that an unknown element, fluorine, was present in the gas, but they were unable to isolate it because of its extreme reactivity.

Fluorine's reactivity makes it a very violent element. For seventy-five years, chemists continually attempted to produce pure fluorine with disastrous results. At least two chemists died while investigating fluorine, and several others narrowly escaped with their lives. Paulin Louyet, a Belgian chemist, died from inhaling hydrogen fluoride vapors. Another scientist, Sir Humphry Davy of England, had to end his attempt to isolate fluorine when he became ill. In 1869, English chemist George Gore was able to produce a small amount of fluorine. However, it reacted violently with the apparatus he was using, causing a great explosion. The French scientist André-Marie Ampère, who had coined the name "fluorine" in 1812, even tried to change its name to phthorine, which was a word derived from the Greek word *phthoros*, meaning "destructive."

Henri Moissan, a French chemist, was well aware of the risks when he began working with fluorine in 1886. He was finally able to successfully

Henri Moissan (1852–1907) was awarded the 1906 Nobel Prize in Chemistry for isolating the element fluorine in 1886 and for creating the Moissan electric-arc furnace in 1892.

isolate pure, or elemental, fluorine by passing electricity through a mixture of hydrogen fluoride (HF) and potassium fluoride (KF) in a process called electrolysis. Moissan solved the problem of fluorine reacting with his equipment by using platinum vessels. Platinum (Pt) is the one element that is able to resist reaction with fluorine for some time. The fluorine reacts with the platinum, forming a skin of platinum fluoride (PtF_4). This skin then protects the platinum from further attack by fluorine. Using this expensive equipment, Moissan became the first to see the pale yellow gas we know as fluorine. For this, he was awarded the Nobel Prize in Chemistry in 1906. A year later, fifty-five-year-old Moissan died, his life undoubtedly cut short by his work with fluorine. Today, fluorine is produced industrially by the same process that Moissan used. It is prepared from fluorspar at a rate of thousands of tons per year. It is transported safely as a liquid in special containers made of nickel (Ni) or steel that are cooled by liquid air.

Arranging the Periodic Table

Fluorine is just one of many elements. Today, scientists know of more than 111 different elements. As scientists discovered more and more elements over the years, they realized that the elements had to be organized somehow. Eventually, the elements were arranged on a big chart, called the periodic table. The periodic table that we use today is based on the work of a Russian chemist named Dmitry Mendeleyev. He published his version of the periodic table in 1869, while teaching chemistry at the

			K = 39	Rb = 85	Cs = 133	—	—
			Ca = 40	Sr = 87	Ba = 137	—	—
			—	?Yt = 88?	?Di = 138?	Er = 178?	—
			Ti = 48?	Zr = 90	Ce = 140?	?La = 180?	Tb = 231
			V = 51	Nb = 94	—	Ta = 182	—
			Cr = 52	Mo = 96	—	W = 184	U = 240
			Mn = 55	—	—	—	—
			Fe = 56	Ru = 104	—	Os = 195?	—
Typische Elemente			Co = 59	Rh = 104	—	Ir = 197	—
			Ni = 59	Pd = 106	—	Pt = 198?	—
H = 1	Li = 7	Na = 23	Cu = 63	Ag = 108	—	Au = 199?	—
	Be = 9,4	Mg = 24	Zn = 65	Cd = 112	—	Hg = 200	—
	B = 11	Al = 27,3	—	In = 113	—	Tl = 204	—
	C = 12	Si = 28	—	Sn = 118	—	Pb = 207	—
	N = 14	P = 31	As = 75	Sb = 122	—	Bi = 208	—
	O = 16	S = 32	Se = 78	Te = 125?	—	—	—
	F = 19	Cl = 35,5	Br = 80	J = 127	—	—	—

In his 1869 periodic table, Dmitry Mendeleyev listed the elements known at that time according to their atomic weights. He left gaps where he believed unknown elements should fit.

University of St. Petersburg in Russia. Mendeleyev sought to organize the elements in a way that would make it easier for his students to study and understand them. He arranged the elements in horizontal rows, according to weight, with the lightest element of each row on the left end and the heaviest on the right, and with one row on top of another so that elements with similar characteristics fell into vertical columns. It is called the "periodic" table because of the way that the chemical behaviors of the elements on the table vary periodically, or at regular, predictable intervals. When Mendeleyev constructed his table, he left gaps in it. He correctly assumed that these gaps belonged to elements that had not yet been discovered. Based on the locations of the gaps, Mendeleyev was able to predict rather accurately the properties, or characteristics, of the yet-to-be discovered elements. Though Mendeleyev's periodic table did not list all of the elements that we know today, fluorine was among those that he included on his first chart.

Chapter Two
The Element Fluorine and the Periodic Table

Although the world is very complex, everything in it—the rocks, the air, and every living thing—is made up of elements. The basic components of each of these elements are atoms. Every element is made up of only one type of atom. Atoms are so tiny that it would take about 100 million fluorine atoms, lying side by side, to form a line that is only 0.4 inch (1 centimeter) long! Amazingly, atoms are made up of even smaller components called subatomic particles. To truly understand what makes elements such as fluorine unique, we have to take a closer look at these particles.

Subatomic Particles

The unique properties of elements are determined by their number of protons, neutrons, and electrons. These are called subatomic particles. Protons and neutrons are clustered together at the center of the atom to form a dense core called the nucleus. Because protons have a positive electrical charge and neutrons carry no electrical charge, the nucleus of an atom has an overall positive electrical charge. Fluorine has nine protons in its nucleus, so its nucleus has a charge of +9.

Electrons are negatively charged particles that are arranged in layers, or shells, around the nucleus of an atom. The electrons are not fixed in

Each fluorine atom has nine protons and ten neutrons in its nucleus. Nine electrons revolve around the nucleus in two shells: two in the innermost shell and seven in the outermost.

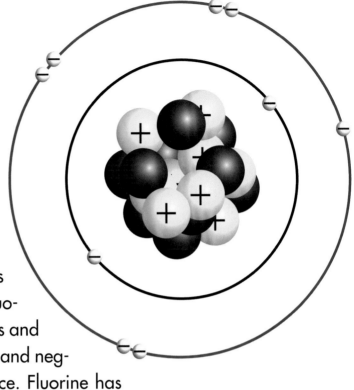

a single position but move rapidly around the nucleus. The negative electrons are attracted to the positive nucleus, and it is this attraction that holds the atom together. In isolated fluorine atoms, the number of protons and electrons is equal, so the positive and negative charges of the atom balance. Fluorine has nine protons in its nucleus and nine electrons outside its nucleus.

Every Element Is Unique

What makes fluorine different from other elements such as oxygen or silver (Ag)? The difference lies in the number of protons that are found in the nuclei of their atoms. The number of protons that are found in the atoms of an element is called the atomic number. On the periodic table reproduced in this book on pages 40–41, the atomic number is found to the upper left of the element's symbol. Because an atom of fluorine has nine protons, its atomic number is nine. The fact that fluorine has nine protons in its nucleus is what makes it fluorine. If you were able to add one proton to fluorine's nucleus, you would end up with an entirely different element called neon (Ne). Neon, which has ten protons in its nucleus, is very nonreactive. What a difference from the extremely reactive fluorine! If you were able to remove one of fluorine's protons, you would have

A Deadly Change

Probably everyone has eaten vinegar sometime in their lives, whether it was sprinkled on french fries or poured over a salad. Vinegar, which is known to chemists as acetic acid (CH_3COOH), contains the elements carbon (C), hydrogen (H), and oxygen. Vinegar is perfectly safe to eat, but if you replace just one of its hydrogen atoms with a fluorine atom, you get fluoroacetic acid (CH_2FCOOH)—something entirely different. You would not want to sprinkle this highly poisonous substance on your french fries. Fluoroacetic acid is found naturally in plants called gifblaar, which are native to South Africa. This chemical is responsible for a number of deaths among the cattle that graze there. Native cultures have even been known to poison the tips of their arrows with it.

oxygen, which has eight protons in its nucleus. We need oxygen to breathe. Just one proton makes the difference between an element we need to breathe and one that would kill us if we breathed it!

Isotopes and Atomic Weight

The number of protons in an atom of a particular element is always the same—that's what makes the element unique. However, many elements exist in different forms called isotopes. Isotopes of an element have the same number of protons and electrons but have a different number of neutrons. Isotopes are commonly referred to by the element's name followed by its mass number. The mass number is the sum of the number of protons and neutrons in the nucleus. For example, carbon-12 is a carbon atom with six protons and six neutrons (6 + 6 = 12), while carbon-14 is a carbon atom with six protons and eight neutrons (6 + 8 = 14). In nature, almost all of the elements exist as a mixture of two or more isotopes. Fluorine has many isotopes, but, unlike most other elements, it has only one natural isotope, called fluorine-19. This isotope has nine protons and ten neutrons (9 + 10 = 19).

The atomic weight of an element is the average weight of all the isotopes of that element, taking into consideration how often each isotope occurs. On the periodic table, the atomic weight of an element is often a number with a decimal fraction. For example, the atomic weight of carbon is 12.011. The atomic weight of fluorine is 18.998404, which is often rounded to 19.

Fluorine and the Periodic Table

Unlike Mendeleyev's chart, the periodic table that we use today lists the elements in order of increasing atomic number (the number of protons). Arranged like this, many trends, or patterns, can be seen. You can use these trends to help classify an element. By seeing where an element is located on the periodic table, you can predict whether it is a metal, a nonmetal, or a metalloid (a substance that has characteristics of both metals and nonmetals).

If you look at the periodic table, you will notice that the elements are divided by a line known as the staircase line. The metals are found to the left of this line and the nonmetals to the right. Most of the elements bordering the line are metalloids. Fluorine is found to the right of the staircase line. This is where you would expect to find it because fluorine is a nonmetallic gas.

By definition, a nonmetal is an element that does not have the characteristics of metals. Metals are easily recognized by their physical traits. Generally, metals are solids that can be polished to be made shiny. They also conduct electricity. Most metals also have the ability to be hammered into shapes without breaking. This is called malleability. Metals are also

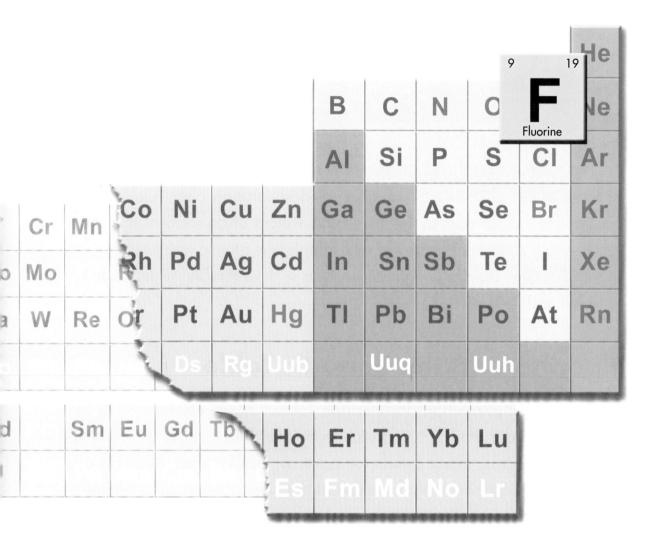

On the periodic table, fluorine heads up the column known as the halogens. In addition to fluorine, the other members of the halogen family are chlorine (Cl), bromine (Br), iodine (I), and astatine (At).

usually ductile, which means that they are able to be pulled into wires. Substances such as wood, glass, or plastics are classified as nonmetals. Unlike metals, nonmetals are not shiny, and they do not conduct heat or electricity well. Nearly half of the nonmetal elements are gases, but some are liquids or solids. However, unlike solid metals, solid nonmetal elements are brittle and will crumble or break apart if pulled on or hammered.

Fluorine 9 F 19 Snapshot

Chemical Symbol:	F
Classification:	Nonmetal; halogen
Properties:	Extremely reactive, pale yellow, corrosive
Discovered By:	Henri Moissan in 1886
Atomic Number:	9
Atomic Weight:	18.998 atomic mass units (amu)
Protons:	9
Electrons:	9
Neutrons:	10
Density at 68°F (20°C):	0.001696 grams per cubic centimeter (g/cm^3)
Melting Point:	–363 °F (–219°C)
Boiling Point:	–307 °F (–188°C)
Commonly Found:	In bones, teeth, blood, urine, saliva, sea water, eggs, hair, and in the earth's crust

Groups and Periods

As you look across the periodic table from left to right, each row of elements is called a period. Elements are arranged in periods by their atomic number, or the number of protons they have in their nuclei. As you go across the table from left to right, the atomic number increases by one from one element to the next. Each period has been assigned a number from one to eight. The period an element is found in can help you to understand how the electrons are arranged in that element. Generally, the number of electron shells that surround the nucleus of an element corresponds to its period number. For instance, fluorine is in period two, so it has two shells of electrons surrounding its nucleus. There are two electrons in fluorine's innermost electron shell and seven in its outermost shell.

As you read down the periodic table from top to bottom, each column of elements is called a group or family. Just as you might have similar characteristics to the other members of your family, the elements in a given family have similar properties. An element's chemical behavior (how it acts) is determined by its electrons, particularly those in its outer shell. The electrons in an element's outermost shell are called valence electrons. All the elements in a given group behave similarly because they all have the same number of electrons in their valence shells.

Fluorine heads up the Group VIIA elements. This is the traditional way to label the groups, using Roman numerals and dividing them into A-groups and B-groups. However, the International Union of Pure and Applied Chemistry (IUPAC), an international group of scientists who set standards in chemistry, have officially adopted a different way to label the groups of the periodic table. They number the groups from left to right using Arabic numerals from 1 to 18. Using the IUPAC system, fluorine's group (Group VIIA) is known as Group 17. Some scientists use the IUPAC system, but most chemists in the United States prefer the more traditional system. The other elements found with fluorine in Group VIIA are chlorine, bromine (Br), iodine (I), and astatine (At). Each of these elements has seven electrons in its outermost electron shell. Together, they form a very special group of elements called the halogens.

Fluorine and the Halogens

The electrons surrounding the nucleus of an atom are arranged in shells. Strict rules determine how many electrons each of these shells is able to hold. For instance, a shell closest to a nucleus is able to hold two electrons, while the next shell out is able to hold eight. Elements are very stable when their outermost electron shells are full. The halogens each have seven electrons in their outermost electron shells. Because their outermost shells can potentially hold eight electrons, they are nearly full. To fill their

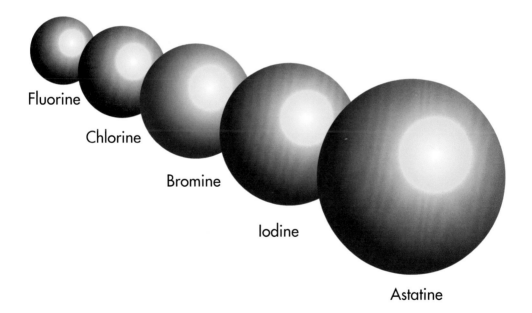

Fluorine

Chlorine

Bromine

Iodine

Astatine

In general, the size of an atom increases as you go down a given group of the periodic table. As one moves down a group, the number of electrons and filled electron shells increases. The fluorine atom is the smallest of the halogens, with nine electrons.

outermost electron shells, the halogens each tend to accept one electron from other elements. Fluorine's atom, with nine electrons, is the smallest of all the halogens. Chlorine has seventeen electrons, bromine has thirty-five, and iodine has fifty-three. The astatine atom is the largest of the halogens, with eighty-five electrons surrounding its nucleus.

The word "halogen" comes from the Greek words *hals*, meaning "salt," and *genes*, meaning "born." The halogens are known as the "salt formers" because of the way they accept electrons from metals to form salts. The halogens are all nonmetals, which means that they are not shiny, do not conduct electricity, and are not ductile or malleable. Of the five elements in the halogen group, only fluorine, chlorine, bromine, and iodine are found in biological molecules. The last and largest halogen, astatine, is radioactively unstable and so is never found in nature. All of the halogens are very reactive.

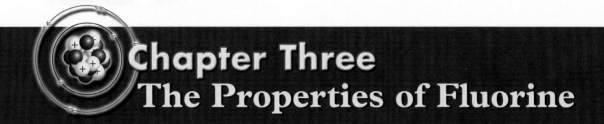

Chapter Three
The Properties of Fluorine

All elements have characteristic physical and chemical properties. These properties help scientists to identify and classify them. The physical properties of an element are those that can be observed without changing the element's identity. Some examples of physical properties are an element's phase (or physical state) at room temperature, density, and freezing point. The chemical properties of an element describe the element's ability to undergo chemical change by combining with other elements. A chemical change converts one kind of matter into a new kind of matter. If an element undergoes chemical change easily, it is said to be very reactive.

Fluorine's Phase at Room Temperature

At room temperature (68°F [20°C]), an element is found in one of three phases: solid, liquid, or gas. Knowing the phase, or physical state, of an element at room temperature helps scientists to identify it. At room temperature, the halogens exist in all three phases. Fluorine is found in the gas phase at room temperature. At room temperature, chlorine is also a gas and is green in color. Bromine is a brown liquid at room temperature, while iodine and astatine are solids. Iodine is purple, but scientists do not know the color of astatine. Because it is extremely unstable, scientists have only been able to artificially produce about one-millionth of a gram of astatine. It is not surprising, therefore, that so very little is known about astatine's properties.

Gases, such as fluorine, do not have definite shapes or volumes. If a gas is put in a container, it will take the shape of that container. Gases differ from solids and liquids in that they are able to be compressed and are able to expand. A gas will fit in a container of almost any size and shape. If it is put in a large container, a gas will expand to fill it. A gas can also be compressed to fit into a smaller container. If not confined to a container, gases will disperse into space.

Cooling Fluorine

At normal room temperature, fluorine is always a gas. However, if it is cooled to a low enough temperature, fluorine will become a liquid or even a solid. For fluorine to condense, or turn into a liquid, the temperature must reach a chilly −307°F (−188°C). At this condensation point (also called boiling point), fluorine turns from a pale yellow gas to a yellow liquid. For fluorine to reach its freezing point or melting point (the temperature at which it becomes a solid), the temperature has to drop even farther, to −363°F (−219°C). As a solid, fluorine is an almost colorless crystal.

As you move down the group in the periodic table, the melting and boiling points of the halogens increase. Fluorine and chlorine have melting and boiling points that are below room temperature. This is why they are found as gases at room temperature. Bromine's boiling point is above room temperature, but its melting point is below room temperature. It is condensed but not frozen. Therefore, bromine is a liquid at room temperature. Iodine has melting and boiling points that are higher than room temperature. For this reason, iodine is still a solid at room temperature.

The Density of Fluorine

Density is another physical property of matter. Density is the amount of a substance contained in a specific volume. Each element has a unique density associated with it. Therefore, an element's density is useful in helping scientists to identify it. Fluorine has a density of 0.001696 grams per cubic centimeter at room temperature. The densities of different gases are often compared to the density of air. If a gas has a lower density than air, it will float in air. In contrast, if a gas has a higher density than air, the gas will sink in air. The density of air is 0.001207 grams per cubic centimeter at room temperature. That makes fluorine about 1.4 times denser than air. Since fluorine is slightly heavier than air, it will gradually sink to the ground if it is released.

A Diatomic Molecule

All of the halogens, except for astatine, are known for forming stable, diatomic molecules. The word "diatomic" means "two atoms." A diatomic

Karl O. Christe

In 1986, Karl O. Christe, while working in a laboratory at Edwards Air Force Base in California, made an astounding discovery. While he was preparing for a conference that would celebrate the 100th anniversary of the discovery of fluorine, he found a way to make fluorine using only chemicals—no electricity. Because of the large amount of energy it takes to separate fluorine from other elements, many scientists had long believed that such a feat was impossible.

23

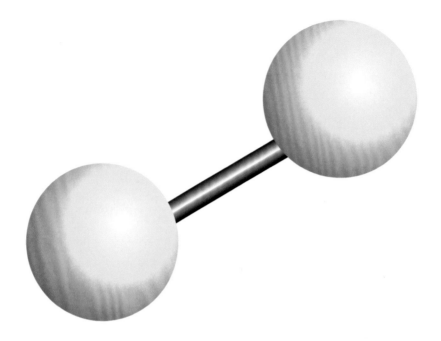

Fluorine gas consists of a pair of fluorine atoms bonded together, forming a diatomic molecule. This type of model of a molecule is called the ball-and-stick model. The two balls represent the two fluorine atoms, while the stick represents the single bond that holds the atoms together.

fluorine molecule contains two fluorine atoms and is written as F_2. In the same way, diatomic chlorine, bromine, and iodine molecules are written as Cl_2, Br_2, and I_2, respectively. Their ability to form diatomic molecules makes the halogens special. Besides the halogens, only three other elements form diatomic molecules like these. They are hydrogen (H_2), oxygen (O_2), and nitrogen (N_2).

In a diatomic fluorine molecule, each fluorine atom shares a pair of electrons with the other atom. When two atoms share electrons in this way, the link they form is called a covalent bond. The diatomic fluorine molecule that results from the covalent bond is much more stable (less reactive) than the individual fluorine atoms. This stability results because, by sharing electrons, each atom is able to fill its outer electron shell.

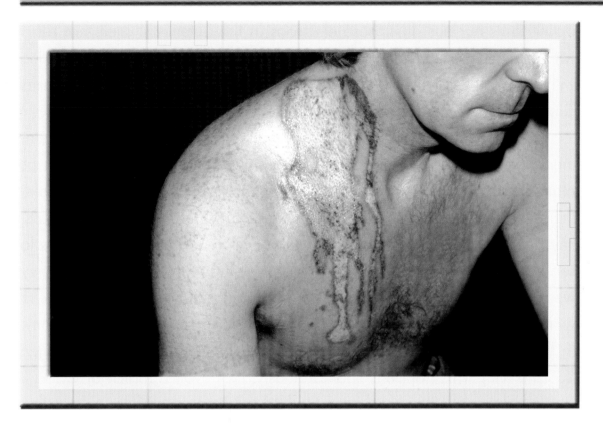

This man's shoulder and chest were burned in an accident when fluorine reacted to the water in his body's cells, forming hydrofluoric acid. Hydrofluoric acid has many industrial uses, including the etching of glass, rust removal and metal cleaning, and the manufacturing of electronics.

Fluorine's Reactivity

Though all of the halogens are reactive, fluorine is by far the most reactive member of the group. This means that it combines with other elements very readily. The key to fluorine's reactivity is the way that it attracts electrons to itself more strongly than any other element. An element that attracts electrons is described as being electronegative. Because the atoms of fluorine and the other halogens need only one more electron in their outermost shell to become stable, they attract electrons very strongly. The fluorine atom is the smallest of the halogens, which means that its outermost electron shell is closer to its nucleus. It is these electrons in the outermost shell that react with other elements. Therefore, the electrons from other elements are pulled with more force toward the positive protons in fluorine's nucleus. This makes fluorine more likely to react with, or combine with, other elements.

Fluorine is so reactive that it combines with almost every other element. It even reacts with some of the most nonreactive elements, such as platinum and gold (Au). Fluorine is unusual because, unlike most elements, it even reacts with the other members of its group, the halogens. Fluorine's extreme reactivity explains why it took so long for scientists to separate it from other elements. It kept reacting with the containers they were trying to collect it in! Fluorine's reactivity also makes it dangerous to work with. When it reacts with other elements, it often does so violently. In the presence of hydrogen, fluorine burns violently, forming hydrogen fluoride (HF). This happens spontaneously. Not even a match or a spark is needed to make it burn! If it is breathed in by accident, fluorine gas can burn the lungs and skin. This burning is caused by fluorine reacting with the water that is found in the body's cells, forming a corrosive chemical called hydrofluoric acid (a solution of HF in water).

Chapter Four
Fluorine Compounds

There are millions of different compounds all around you. A compound is formed when two or more elements are bonded together. Fluorine is so reactive that it bonds with virtually every element. For this reason, fluorine is never found by itself in nature. It is always found in compounds. During a reaction, fluorine steals electrons from other elements to fill its outer electron shell.

All of the halogens have a tendency to gain one electron, bringing the number of electrons in their outermost shells to eight. Atoms are usually electrically neutral, which means they carry no charge. They carry no charge because they have an equal number of positively charged protons and negatively charged electrons. However, if an atom picks up extra negatively charged electrons, it becomes a negatively charged ion. In the same manner, if an atom loses electrons, it becomes a positively charged ion. When a fluorine atom gains an electron, it becomes a fluoride ion. Fluoride ions have a −1 charge because they have one extra electron in the outer shell.

Once fluorine has filled its outer electron shell, it will not give up the electron easily. For this reason, the compounds that fluorine forms are very stable. Fluorine bonds with other elements to form many different, interesting, and useful compounds.

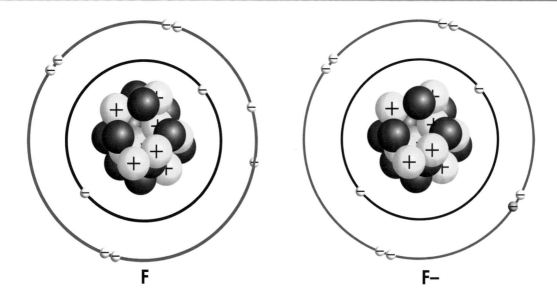

F   F–

There are two electrons in the inner electron shell and seven electrons in the outer shell of a neutral fluorine atom *(left)*. However, a fluorine atom is stable only when it has eight electrons in its outer shell. Fluorine seeks to complete its outer shell by gaining an extra electron, which creates a negatively charged fluoride ion *(right)*.

Hydrofluoric Acid

Hydrogen fluoride (HF) gas dissolves in water to form hydrofluoric acid. Hydrofluoric acid is extremely corrosive. Because it is so destructive, it must be stored in lead (Pb), steel, or plastic containers (that are themselves made with fluorine). Hydrofluoric acid even dissolves glass. Around 1670, the use of hydrofluoric acid for etching glass was first described by Heinrich Schwanhard. Today, this fluorine compound is still used to etch glass.

During etching, glass is first covered by an acid-resistant substance, such as wax, to protect the parts of the glass outside the pattern. The acid then eats away at the uncovered surface of the glass. An etched glass surface may take on a rough, frosted appearance or have a soft, satiny appearance. Etching is used to create designs in glassware or ceramics. It is also used to mark the divisions on thermometer tubes and to create the frosty, satiny finish on the inside of electric lightbulbs.

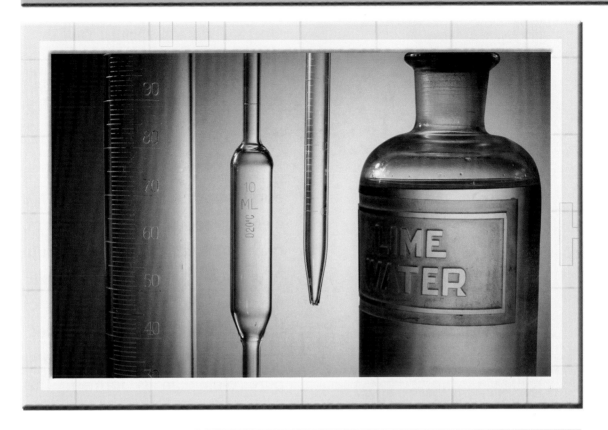

Glass etching uses acid to create patterns on the surface of glass. The graduated markings and labels on this laboratory glassware were created using hydrofluoric acid.

Sodium Aluminum Fluoride

For thousands of years, people have been using minerals containing fluorine to purify metal. They used these minerals without even knowing that fluorine existed. We still use fluorine compounds to purify metals. In fact, the largest use of fluorine compounds today is for the refining of aluminum (Al) metal. Like fluorine, aluminum is always found in combination with other elements in nature. To obtain the pure aluminum that we use to make soda cans, foil, cars, and airplanes, aluminum must first be separated from the other elements to which it is bonded. To separate aluminum from other elements, a fluorine compound called cryolite is needed. Cryolite, or sodium aluminum fluoride (Na_3AlF_6), is a mineral found in the

In 1886, French scientist Paul Heroult (1863–1914) discovered that aluminum oxide would dissolve in cryolite. It could be broken down by electricity into crude, molten aluminum metal.

earth's crust. It is composed of the elements sodium (Na), aluminum, and fluorine. Impure aluminum (aluminum bound to other elements) is dissolved in a bath of hot, molten cryolite. When electricity is passed through the bath, pure aluminum is formed. This process is called electrolysis.

Uranium Hexafluoride

Uranium hexafluoride (UF_6) is a fluorine-containing compound that is produced from fluorine gas. This gaseous compound is used to refine the uranium fuel that is used in nuclear power plants. Nuclear power plants make electricity using the energy produced by the fission of uranium fuel. During fission, uranium splits into two lighter elements. This process gives off an extreme amount of energy in the form of heat and radiation.

Uranium exists as many different isotopes. The isotopes of uranium have different masses (the sum of the protons and neutrons in the nucleus). Pure uranium consists of mostly U-238 (uranium with a mass of 238). However, only the U-235 (uranium with a mass of 235) isotope can be used as nuclear fuel. Fluorine helps to separate these two isotopes so that the U-235 isotope can be used. Pure uranium is combined with fluorine to form uranium hexafluoride (UF_6), a rather exotic and remarkable gas. Its molecules are extremely heavy for a gas—most gases have much lighter molecules. This gas is spun at high speeds and filtered through a thin membrane to obtain the useful U-235.

The Montreal Protocol

The Montreal Protocol is an international treaty that was designed to protect the ozone layer. It phased out the production of the substances believed to be responsible for the depletion of the ozone layer. The treaty called for the eventual end of all manufacture of CFCs. The Montreal Protocol took effect on January 1, 1989. Because of its acceptance by so many nations (as of 2006, 189 nations had ratified the protocol), the Montreal Protocol has been hailed as an exceptional example of international cooperation.

Chlorofluorocarbons

The chlorofluorocarbons, or CFCs, are a group of compounds that contain the elements fluorine, chlorine, and carbon. The chlorofluorocarbons are odorless, nonflammable, noncorrosive, nontoxic chemicals. CFCs were developed in the late 1920s. They soon became widely used as refrigerants in refrigerators, freezers, air conditioners, and heat pumps. They were also used as propellants in aerosol dispensers. They also served as insulating foam in packaging materials, car seats, and bedding.

By the 1970s, however, scientists were concerned that CFCs escaping into the atmosphere were damaging the ozone layer in the upper atmosphere. Ozone is formed in the stratosphere, the layer above the air we

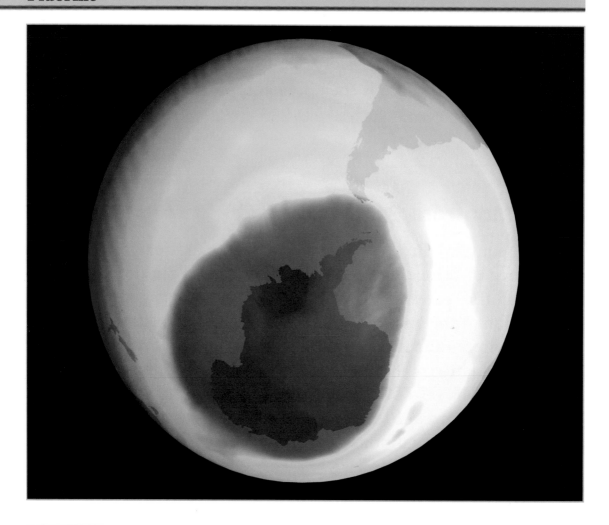

The dark blue area in this picture shows a "hole" in the ozone layer above Antarctica that is three times larger in area than the United States. Scientists detected the harmful effects of CFCs on the ozone layer in the 1970s.

breathe, by the action of ultraviolet light on oxygen. Ozone gas is a special oxygen molecule that is made up of three oxygen atoms (O_3). The ozone layer helps us by blocking out the rays of the sun that can harm our skin and the tissues of plants—humans couldn't last long without plants. If the protective ozone layer were to disappear, we would be exposed to dangerous levels of ultraviolet radiation. This could weaken our immune systems or cause cataracts or cancer. Because of their effects on the ozone layer, the United States banned the use of CFCs in aerosol dispensers in 1978.

Chapter Five
Fluorine and You

When elements join together to form compounds, they lose their individual traits. When fluorine forms compounds with other elements, it undergoes interesting and often dramatic changes to its properties. Though fluorine can be a very dangerous element, its compounds can be very helpful to us.

Fluorine and Your Teeth

Tooth decay from eating too much sugar has become a real problem for people in many countries. One of the remarkable uses of fluorine is to keep teeth from developing cavities. Teeth are protected by a hard enamel made of a mineral called hydroxyapatite. In the presence of sugar, bacteria in the mouth form acids that break down the hydroxyapatite. This loss of tooth enamel causes teeth to decay.

Toothpastes, which contain fluorine in the form of sodium fluoride (or sodium fluorophosphates), can help to prevent cavities. In humans, fluoride accumulates in the bones and teeth. When sodium fluoride is present, the body makes a modified form of tooth enamel. Instead of hydroxyapatite, it forms fluorapatite. Fluorapatite is more resistant to attack from acids in the mouth. Because of this, many communities in the United States now add fluoride to the public drinking water.

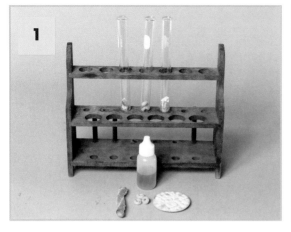

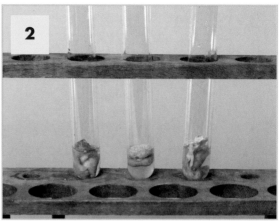

The amount of sugar in the food we eat is a major factor in the development of tooth decay. A pretzel, Honey Nut Cheerios, and a Carr's cracker were crushed and tested with Benedict's solution *(blue liquid)* to determine sugar content. In the presence of some sugars, a green, brown, or red colored precipitate will form. All three foods contain sugar.

Fluorine Kills

Most doctors and dentists agree that a small amount of fluoride is good for maintaining healthy teeth. However, fluoride is believed to be safe only in tiny doses. In large amounts, fluoride is highly dangerous. One one-hundredth of an ounce (0.25 gram) is toxic to humans, and 0.18 ounce (5 grams) could be fatal. In 1943, in a hospital in the United States, patients were served scrambled eggs to which sodium fluoride roach powder had been added instead of powdered milk. Of the patients that ate the poisoned eggs, 263 became ill and 47 died. Because of its ability to kill, sodium fluoride is used as an effective insecticide for cockroaches and ants.

Fluorine and Food

The average person consumes between 0.000011 and 0.00011 ounces (0.3–3 milligrams) of fluoride every day. Most of this fluoride is provided by the fluoridated water supply, but fluoride is found in other foods as well. Foods such as pork, eggs, chicken, potatoes, cheese, and butter all contain traces of fluoride. Even a cup of tea can provide you with as much as 0.000014 ounces (0.4 mg) of fluoride. Sea water contains a good amount of fluoride, too. Though we don't drink sea water, we eat foods that come from the sea. Because of the environment from which they come, things like cod, salmon, mackerel,

The fluoride content of most foods is low. However, some foods, such as salmon, mackerel, sardines, and chicken, are rich sources of fluoride. Fish that is consumed with its bones, such as sardines and mackerel, provides more fluorine than deboned fish.

Fluorosis

Fluorosis is caused by the consumption of too much fluoride. Fluoride can enter the body through toothpaste, drinking water, drugs, and fluoride dust and fumes from industries using fluoride-containing salt or hydrofluoric acid. Fluoride that has been consumed then accumulates in the bones and teeth. There are two types of fluorosis: skeletal and dental fluorosis. Skeletal fluorosis is a crippling bone disease that mostly affects people in India and China. In India, the most common cause of fluorosis is naturally fluoridated water found in wells dug deep into the earth. In China, most people become sick by burning fluoride-enriched coal in their homes for fuel. The early symptoms of skeletal fluorosis are similar to those of arthritis, including painful, stiff joints; the sensation of burning or tingling in the limbs; and muscle weakness. However, the end result can be quite disfiguring. Arms and legs become weak, it becomes difficult to move joints, and the vertebrae partially fuse together, crippling the patient. Dental fluorosis is characterized by white specks or brown markings on the teeth. It has been found to affect from 30 percent to 50 percent of children in areas where fluoride has been added to the drinking water.

and sardines are particularly rich in fluoride. Even vegetables contain a little bit of fluoride.

Cookware, Raincoats, and Hearts

One of fluorine's compounds, tetrafluoroethylene (C_2F_4), has proven to be very useful to us. It is made from an extract of crude oil (ethylene)

This experiment shows how fluoride rinse can protect teeth. Egg shells, like teeth, can be weakened by acid. Photo 1 shows an egg being placed in a fluoride rinse solution. Photo 2 shows the fluoride-treated egg being placed in a cup of white vinegar (which is a weak acid, similar to those found in a person's mouth). An untreated egg is also placed in vinegar. In photo 3, the untreated egg *(right)* bubbles as the acid attacks its shell, but the treated egg *(left)* is protected from attack.

and fluorine. When tetrafluoroethylene molecules are linked together to form a chain, they are known as polytetrafluoroethylene, or Teflon. This is a soft, white plastic that virtually nothing sticks to. Teflon is the nonstick surface used on many household cooking utensils and cookware. It helps to make meals healthier because it reduces the need for fat in cooking. Teflon also makes cleaning up after dinner a lot easier because food does not burn and stick to the bottom of Teflon-coated pots and pans.

Because polytetrafluoroethylene has such amazing properties, it has many other uses as well. In 1969, Bob Gore, chemical engineer and cofounder of DuPont, discovered a way to expand the compound so that it formed invisible pores. He called the material Gortex. Gortex is considered waterproof because its pores are small enough to keep liquid water droplets out, but large enough to allow the gaseous water molecules from sweat to escape. Because of this property, it is used to make rain gear. Gortex is not only good for a rainy day, but it is also used to create artificial veins and arteries for people with cardiovascular disorders. Considering the extreme reactivity of fluorine gas, it may seem surprising that fluorine's compounds are stable enough for one of them to be used as part of someone's heart. However, fluorine compounds are perfectly safe. It is fluorine's extreme reactivity that makes its compounds, such as Teflon and Gortex, so stable. Once a reactive

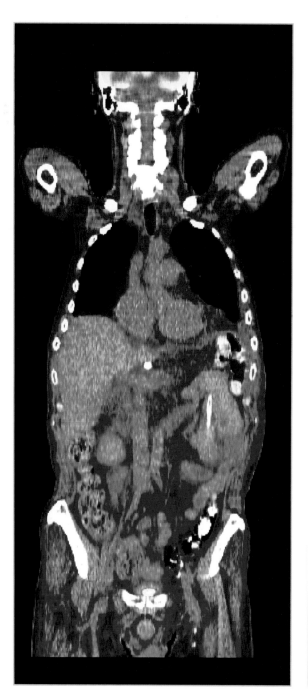

Fluorine-18 is ideal for use in PET scans because it does not remain in the body very long. This minimizes the injury to the patient's body tissues by radiation. This PET scan was done on a patient whom doctors are treating for lung cancer.

fluorine atom has stolen an electron from another atom to form a compound, it is not likely to give up that electron. This makes Gortex a perfectly reasonable choice for someone's heart.

Fluorine and Medicine

Though fluorine has many isotopes, there is only one naturally occurring isotope of fluorine, fluorine-19. The rest of fluorine's isotopes are radioactive and artificially produced. One of these radioactive isotopes, fluorine-18, is used in medical diagnosis. It is used in a medical procedure called positron-emission tomography, or PET. During the procedure, a sample of fluorine-18 is introduced into the body. As it decays, it emits X-ray-like radiation. This radiation can be scanned by special instruments to reveal a picture of the body's vital organs.

Fluorine compounds are so varied and affect our lives in so many different ways. Fluorine killed and injured those who sought to discover it, yet it is used in artificial hearts to save lives. Fluorine compounds are employed in pesticides, but they are also added to drinking water and toothpastes to help keep your teeth healthy. Fluorine compounds are found not only in uranium and aluminum refineries, but also in your kitchen. Indeed, life would be a lot more difficult and a lot less interesting if it were not for the element fluorine!

The Periodic Table of Elements

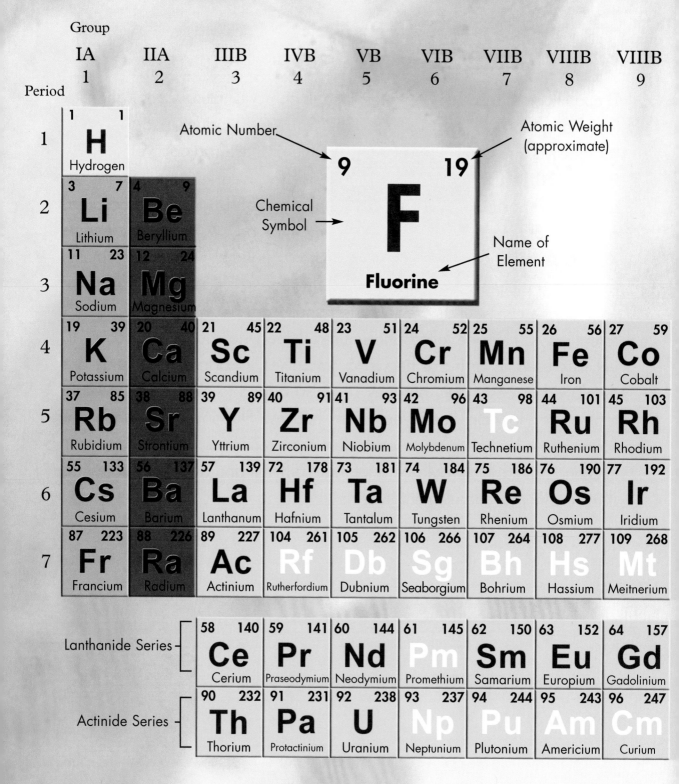

Group

	IA	IIA	IIIB	IVB	VB	VIB	VIIB	VIIIB	VIIIB
	1	2	3	4	5	6	7	8	9

Period

Atomic Number
Atomic Weight (approximate)

9   19
F
Fluorine

Chemical Symbol

Name of Element

1 — 1 1 **H** Hydrogen

2 — 3 7 **Li** Lithium | 4 9 **Be** Beryllium

3 — 11 23 **Na** Sodium | 12 24 **Mg** Magnesium

4 — 19 39 **K** Potassium | 20 40 **Ca** Calcium | 21 45 **Sc** Scandium | 22 48 **Ti** Titanium | 23 51 **V** Vanadium | 24 52 **Cr** Chromium | 25 55 **Mn** Manganese | 26 56 **Fe** Iron | 27 59 **Co** Cobalt

5 — 37 85 **Rb** Rubidium | 38 88 **Sr** Strontium | 39 89 **Y** Yttrium | 40 91 **Zr** Zirconium | 41 93 **Nb** Niobium | 42 96 **Mo** Molybdenum | 43 98 **Tc** Technetium | 44 101 **Ru** Ruthenium | 45 103 **Rh** Rhodium

6 — 55 133 **Cs** Cesium | 56 137 **Ba** Barium | 57 139 **La** Lanthanum | 72 178 **Hf** Hafnium | 73 181 **Ta** Tantalum | 74 184 **W** Tungsten | 75 186 **Re** Rhenium | 76 190 **Os** Osmium | 77 192 **Ir** Iridium

7 — 87 223 **Fr** Francium | 88 226 **Ra** Radium | 89 227 **Ac** Actinium | 104 261 **Rf** Rutherfordium | 105 262 **Db** Dubnium | 106 266 **Sg** Seaborgium | 107 264 **Bh** Bohrium | 108 277 **Hs** Hassium | 109 268 **Mt** Meitnerium

Lanthanide Series — 58 140 **Ce** Cerium | 59 141 **Pr** Praseodymium | 60 144 **Nd** Neodymium | 61 145 **Pm** Promethium | 62 150 **Sm** Samarium | 63 152 **Eu** Europium | 64 157 **Gd** Gadolinium

Actinide Series — 90 232 **Th** Thorium | 91 231 **Pa** Protactinium | 92 238 **U** Uranium | 93 237 **Np** Neptunium | 94 244 **Pu** Plutonium | 95 243 **Am** Americium | 96 247 **Cm** Curium

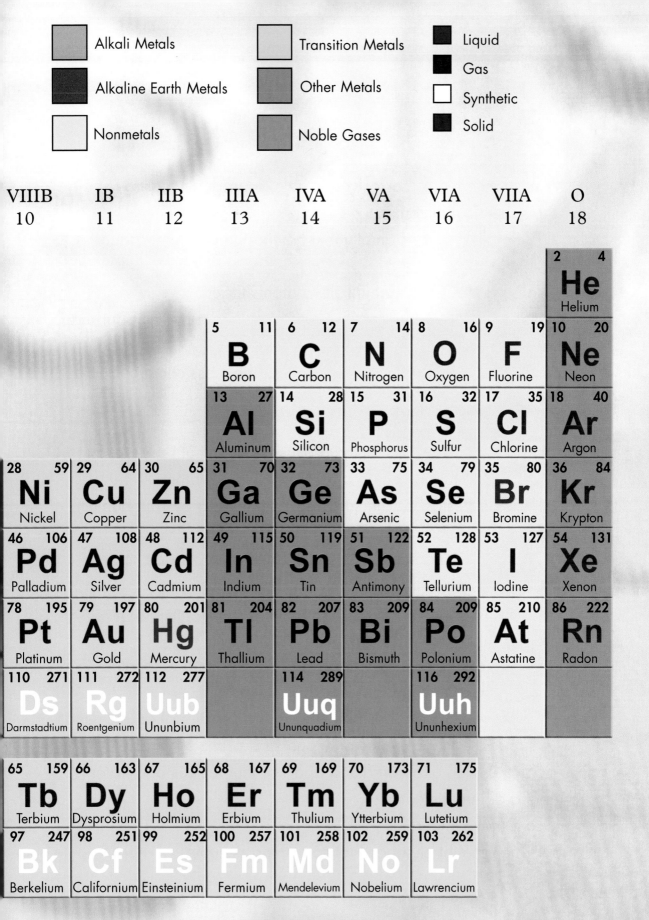

Legend

- Alkali Metals
- Alkaline Earth Metals
- Nonmetals
- Transition Metals
- Other Metals
- Noble Gases
- Liquid
- Gas
- Synthetic
- Solid

VIIIB 10	IB 11	IIB 12	IIIA 13	IVA 14	VA 15	VIA 16	VIIA 17	O 18
								2 4 **He** Helium
			5 11 **B** Boron	6 12 **C** Carbon	7 14 **N** Nitrogen	8 16 **O** Oxygen	9 19 **F** Fluorine	10 20 **Ne** Neon
			13 27 **Al** Aluminum	14 28 **Si** Silicon	15 31 **P** Phosphorus	16 32 **S** Sulfur	17 35 **Cl** Chlorine	18 40 **Ar** Argon
28 59 **Ni** Nickel	29 64 **Cu** Copper	30 65 **Zn** Zinc	31 70 **Ga** Gallium	32 73 **Ge** Germanium	33 75 **As** Arsenic	34 79 **Se** Selenium	35 80 **Br** Bromine	36 84 **Kr** Krypton
46 106 **Pd** Palladium	47 108 **Ag** Silver	48 112 **Cd** Cadmium	49 115 **In** Indium	50 119 **Sn** Tin	51 122 **Sb** Antimony	52 128 **Te** Tellurium	53 127 **I** Iodine	54 131 **Xe** Xenon
78 195 **Pt** Platinum	79 197 **Au** Gold	80 201 **Hg** Mercury	81 204 **Tl** Thallium	82 207 **Pb** Lead	83 209 **Bi** Bismuth	84 209 **Po** Polonium	85 210 **At** Astatine	86 222 **Rn** Radon
110 271 **Ds** Darmstadtium	111 272 **Rg** Roentgenium	112 277 **Uub** Ununbium		114 289 **Uuq** Ununquadium		116 292 **Uuh** Ununhexium		

65 159 **Tb** Terbium	66 163 **Dy** Dysprosium	67 165 **Ho** Holmium	68 167 **Er** Erbium	69 169 **Tm** Thulium	70 173 **Yb** Ytterbium	71 175 **Lu** Lutetium
97 247 **Bk** Berkelium	98 251 **Cf** Californium	99 252 **Es** Einsteinium	100 257 **Fm** Fermium	101 258 **Md** Mendelevium	102 259 **No** Nobelium	103 262 **Lr** Lawrencium

acid A substance that is corrosive, or able to gradually destroy something by chemical action.

atom The smallest part of an element having the chemical properties of that element.

bond An attractive force that links two atoms together.

chemical reaction A change in which one kind of matter is turned into another kind of matter.

density A physical property of an element; the mass contained in a unit of volume.

dissolve To be absorbed into a liquid.

electrolysis The use of electricity to cause a chemical reaction.

element A substance made up of only one type of atom.

etch To make a design on the surface of something by roughening the surface.

halogen An active nonmetal in Group VIIA of the periodic table.

impurity A substance that contaminates something.

isotopes Atoms containing the same number of protons and electrons but different numbers of neutrons.

molecule A group of atoms that are chemically bonded together.

molten Melted, or changed into a liquid using heat.

property (of an element) Describes the characteristics of an element.

pure A substance that contains only one kind of molecule or atom.

radioactive decay The breaking apart of the nucleus of an atom, forming a different element.

refine To remove impurities.

volume The amount of space that something occupies.

For More Information

American Chemical Society—Division of Fluorine Chemistry
1155 Sixteenth Street NW
Washington, DC 20036
(800) 227-5558
Web site: http://www.chemistry.org

American Dental Association
211 East Chicago Avenue
Chicago, IL 60611-2678
(312) 440-2500
Web site: http://www.ada.org

Fluoride Action Network
P.O. Box 5111
Burlington, VT 05402
(802) 355-0999
(315) 379-9200
Web site: htttp://www.fluoridealert.org

National Center for Fluoridation
410 North Michigan Avenue, Suite 352
Chicago, IL 60611
(312) 836-9900
Web site: http://www.fluoridationcenter.org

New York State Coalition Opposed to Fluoridation (NYSCOF)
P.O. Box 263
Old Bethpage, NY 11804
Web site: http://www.orgsites.com/ny/nyscof

Web Sites

Due to the changing nature of Internet links, Rosen Publishing has developed an online list of Web sites related to the subject of this book. This site is updated regularly. Please use this link to access the list:

http://www.rosenlinks.com/uept/fluo

Curran, Greg. *Homework Helpers: Chemistry*. Franklin Lakes, NJ: Career Press, 2004.

Greenwood, N. N., and A. Earnshaw. *Chemistry of Elements*. Oxford, England: Pergamon Press, 1984.

Jackson, Tom. *The Elements: Fluorine*. New York, NY: Benchmark Books, 2004.

Oxlade, Chris. *Elements and Compounds* (Chemicals in Action). Chicago, IL: Heinemann Library, 2002.

Saunders, Nigel. *Fluorine and the Halogens* (The Periodic Table). Chicago, IL: Heinemann, 2004.

Stwertka, Albert. *A Guide to the Elements*. 2nd ed. New York, NY: Oxford University Press, 2002.

Bibliography

Brady, James E., and John R. Holum. *Chemistry: The Study of Matter and Its Changes.* New York, NY: John Wiley & Sons, Inc., 1993.

Ebbing, Darrell D. *General Chemistry.* 4th ed. Boston, MA: Houghton Mifflin Company, 1993.

Emsley, John. *Nature's Building Blocks.* New York, NY: Oxford University Press, 2001.

Heiserman, David L. *Exploring Chemical Elements and Their Compounds.* New York, NY: Tab Books, 1992.

Newton, David E. *Chemical Elements: From Carbon to Krypton.* Farmington Hills, MI: The Gale Group, 1999.

Stwertka, Albert. *A Guide to the Elements.* 2nd ed. New York, NY: Oxford University Press, 2002.

Index

About the Author

Heather Hasan graduated summa cum laude from East Stroudsburg University with a dual major in biochemistry and chemistry. She loves science and has written about other elements of the periodic table, including iron, nitrogen, helium, and aluminum. Ms. Hasan currently lives in Greencastle, Pennsylvania, with her husband, Omar, and their sons, Samuel and Matthew.

Photo Credits

Cover, pp. 1, 13, 16, 19, 24, 28, 40–41 by Tahara Anderson; p. 5 courtesy of Hagley Museum and Library; p. 7 © José Manuel Sanchis Clavete/Corbis; pp. 9, 30 © Bettmann/Corbis; p. 10 © SPL/Photo Researchers, Inc.; p. 25 © 2000–2006 Custom Medical Stock Photo; p. 29 © Andrew Lambert Photography/Photo Researchers, Inc.; p. 32 © NASA Goddard Space Flight Center; pp. 34, 37 by Mark Gobeliowski; p. 35 (top left) © 2006 iStock International Inc.; p. 35 (top right) © Mark Huntington; p. 35 (lower left) © Geralda van der Es; p. 35 (lower right) © Paul Johnson; p. 38 © Beil Borden/Photo Researchers, Inc.

Special thanks to Jenny Ingber, high school chemistry teacher, Region 9 Schools, New York City, New York, for her assistance in executing the science experiments in this book.

Designer: Tahara Anderson; Editor: Kathy Kuhtz Campbell

Puedes consultar nuestro catálogo en www.picarona.net

El secreto del tonto del pueblo
Texto: *Rebecca Upjohn*
Ilustraciones: *Renné Benoit*

1.ª edición: febrero de 2017

Título original: *The Secret of the Village Fool*

Traducción: *Oriana Bonet*
Maquetación: *Isabel Estrada*
Corrección: *M.ª Ángeles Olivera*

© 2012, Rebecca Upjohn para el texto
© 2012, Renné Benoit para las ilustraciones
(Reservados todos los derechos)

Publicado con permiso de Second Story Press, Toronto, Canadá.

Fotos de las páginas 30, 34, 35 y 36 usadas con permiso © colección de la familia Adler y de Yad Vashem.
Fotos de las páginas 31, 32, 33, 34 y 36 cortesía de la familia Zeiger.

© 2017, Ediciones Obelisco, S. L.
www.edicionesobelisco.com
(Reservados los derechos para la lengua española)

Edita: Picarona, sello infantil de Ediciones Obelisco, S. L.
Collita, 23-25. Pol. Ind. Molí de la Bastida
08191 Rubí - Barcelona - España
Tel. 93 309 85 25 - Fax 93 309 85 23
E-mail: picarona@picarona.net

ISBN: 978-84-9145-023-8
Depósito Legal: B-98-2017

Printed in Spain

Impreso en España por ANMAN, Gràfiques del Vallès, S. L.
C/ Llobateres, 16-18, Tallers 7 - Nau 10. Polígon Industrial Santiga.
08210 - Barberà del Vallès (Barcelona)

El secreto *del tonto* del pueblo

Texto: Rebecca Upjohn
Ilustraciones: Renné Benoit

Milek y su hermano, Munio, vivían en un tranquilo y pequeño pueblo de Polonia.

Un día de primavera, su madre les entregó un cazo lleno de sopa, una hogaza de pan y unas cuantas prendas viejas.

—Llevad esto a casa de Anton Suchinski –dijo.

—No quiero ir, mamá. Anton es raro. Habla con los animales –contestó Milek.

—Habla con los animales y con las plantas –añadió Munio, que era tres años mayor y sabía más de todo.

Pero la expresión de mamá decía que iban a ir, tanto si les gustaba como si no. Se puso las manos en las caderas.

—Puede que Anton tenga su manera especial de hacer las cosas, pero es un buen hombre. Y en estos tiempos pasa hambre. –Los hizo salir–. Decidle que la sopa es sólo de verduras, y regresad enseguida a casa. No quiero que andéis paseando por el pueblo.

Cuando los hermanos llegaron a la casa de Anton, lo encontraron cuidando el jardín. Sonrió en cuanto los vio.

—¡Oh, queridos, que alegría veros!

Era sólo un poco más alto que Munio. Sus ropas remendadas olían a tierra y sus ojos eran tan vivos como los de un pájaro.

—Mamá te manda sopa y pan dijo Munio—. Y esto —añadió, mostrándole dos camisas viejas de su padre y una chaqueta.

Milek escuchaba a Munio y no decía nada. Había oído muchas historias raras acerca de ese hombre. Anton no sabía leer ni escribir. Daba las gracias al sol y a la lluvia porque hacían que crecieran las plantas. Decían que nunca comía carne. Y también que ponía platos con agua azucarada para alimentar a las moscas, como si no hubiera suficientes moscas ya. La gente lo llamaba el tonto del pueblo.

—Mamá ha dicho que la sopa sólo es de verduras —aclaró Munio.

—¡Qué amable es vuestra madre! —contestó Anton—. Y el pan huele muy bien. Por favor dadle las gracias de mi parte.

—¿Por qué no comes carne? —preguntó Milek. Munio se quedó mirándole. Pero Milek quería saber si las historias sobre Anton eran ciertas.

Los ojos de Anton se empequeñecieron mientras se agrandaba su sonrisa.

—¿A vosotros os gustaría ser la cena de alguien? Pues a los animales tampoco. La vida tiene un gran valor. Nadie debería quitársela a nadie.

Milek no respondió.

—Tenéis un largo camino de regreso a casa. ¿Queréis compartir conmigo este festín?

A Milek le gustó la idea, pero Munio movió la cabeza en señal de negación.

—No podemos quedarnos. Mamá se preocuparía.

Mientras los niños se despedían, un vecino que pasaba cerca del patio se detuvo y los miró. Munio tiró de Milek. Conocían a ese hombre. *Tata,* su padre, les había advertido de que debían alejarse de él.

—Cuidado con los amigos que escoges, Suchinski –gruñó el hombre–. Esos chicos son judíos. Yo de ti me alejaría de ellos.

A Anton se le borró la sonrisa de la cara.

—Eso que dices no está bien.

—Eres realmente tonto, ¿no? Tendrás una sorpresa. ¡Tú y tus ocurrencias! Se avecina la guerra. Cuando lleguen los ejércitos de Hitler, esos judíos tendrán problemas. Y sus amigos también, ¡ya lo verás!– Se rio y se marchó.

Milek sintió un nudo en el estómago. ¿Qué quiere decir? ¿Qué guerra? ¿Quién es Hitler?

Munio tomó a Milek del brazo.

—Tenemos que irnos a casa. ¡Ahora!

—Iré con vosotros –dijo Anton– visiblemente preocupado.

Después de la sabrosa cena de sopa con pan, Anton miró la puesta de sol y pensó en sus jóvenes amigos. Adolf Hitler, el dirigente de Alemania, quería conquistar toda Europa. Hitler y sus soldados nazis odiaban a los judíos. ¿Llegarían hasta ese pueblo? Y si era así, ¿qué les pasaría a Milek y a Munio y a su familia, y a los demás judíos? ¿Quién los ayudaría?

Se sentó en su silla de pensar hasta muy tarde, preocupado, muy preocupado.

En verano la guerra llegó al pueblo. Las explosiones retronaban a lo lejos y los aviones de guerra cruzaban el cielo. Las bombas caían por doquier. Llegaron los nazis. Columnas de soldados desfilaron con tanques y armas pesadas, y tomaron posesión del lugar. Poco después empezaron a arrestar a los judíos.

Una mañana, temprano, a Milek le despertó alguien que sollozaba en la habitación de al lado. Él y Munio salieron sigilosamente y se encontraron a mamá sentada rodeando con sus brazos a Eva, una amiga de la familia. A Eva se le caían las lágrimas mientras hablaba.

—¡Ya no están! ¡Toda mi familia! Mi madre hizo que me marchara para esconderme de los soldados. Cuando regresé, nuestra casa estaba vacía. Ya no había nadie.

—Las cosas están muy mal –murmuró *Tata*–. Los nazis han ordenado que todos los judíos del pueblo vayan a la sinagoga. Hay rumores de que los soldados se van a llevar a todos los muchachos.

Mamá se quedó sin aliento. Munio miraba fijamente a *Tata* sin dar crédito a sus oídos.

—No se nos pueden llevar. ¡No pueden! No iré.

—Mamá –sollozó Milek corriendo hacia ella–. ¡Quiero quedarme contigo!

Pom, pom, pom. Todos lo oyeron. Alguien llamaba a la puerta de atrás.

—Finjamos que no hay nadie en casa –susurró mamá, pero *Tata* ya había abierto la puerta.

Allí estaba Anton, con un fardo de ropa.

—Oh, queridos, menos mal que aún estáis aquí –dijo al entrar–. Temo por vosotros. Tengo un plan para salvaros. Pero tenéis que salir esta misma noche. –Vio a Eva–. Tú también tienes que venir –añadió.

Todos se lo quedaron mirando. Anton, el raro, el pobre Anton. Anton, que daba de comer a las moscas, se ofrecía para salvarlos. ¿Qué podía hacer él?

Tata sacudió la cabeza.

—Te lo agradecemos, Anton, pero no podemos aceptar. Si los nazis descubren que nos estás ayudando… te matarán.

Anton sonrió.

—¿Por qué tendrían que molestarse conmigo? Sólo soy el tonto del pueblo, todos lo creen así. Pero tengo un plan. Tenéis que confiar en mí.

Tata dudó unos instantes y luego asintió.

—De acuerdo, Anton. Gracias.

—Venid a mi casa esta noche. –Le dio la ropa a mamá–. Para los chicos; es un disfraz.

Salió mientras mamá sacaba la ropa: dos vestidos y dos pañuelos.

—¡No me voy a poner *eso*! Es ropa de niña –Milek graznó.

—Esa ropa te puede salvar la vida –susurró Eva.

Munio entendió de repente lo que quería decir y agarró un vestido.

—Vamos, Milek –le animó mientras mamá le ayudaba a ponérselo. Ceñudo, Milek dejó que Eva lo vistiera.

Oyeron disparos fuera. Munio corrió hacia la ventana y miró.

—Los soldados obligan a las familias judías a que salgan de sus casas –dijo.

La mano de mamá tembló mientras anudaba el pañuelo que cubría los cabellos de Milek. Munio intentaba atarse el suyo, pero Eva tuvo que enseñarle cómo hacerlo.

—Ven –dijo *Tata* a mamá–. Llevaremos a nuestras niñas a la sinagoga. –Puso la mano en el hombro de Eva–. A las tres.

Cuando llegaron a la sinagoga, encontraron a las otras familias judías apiñadas fuera. Los soldados sacaban los rollos sagrados de la Torá y los quemaban. Después prendieron fuego a la sinagoga.

Mamá abrazaba a Milek y a Munio. Milek quería gritar: «¿Por qué nos odiáis tanto?». Pegaron a un hombre que quería marcharse con su familia. Los soldados dispararon al aire y amenazaron con disparar a todo el que se negara a quedarse mirando cómo ardía la sinagoga.

Luego, los nazis arrancaron a los niños varones de sus familias y los obligaron a ponerse en fila. Milek y Munio estaban de pie con Eva, con las niñas. ¿Engañarían los vestidos a los soldados? El corazón de Milek latía tan fuerte que temía que lo descubrieran. Cerró los ojos y sintió cómo su hermano le tomaba de la mano. Munio también temblaba.

Los soldados se alejaron con los muchachos, sin atender a las súplicas y a los lamentos de sus familias. ¿Cómo podía pasar algo semejante aquí, en su propio pueblo?

—Tan pronto como anochezca —susurró *Tata*— iremos a casa de Anton.

Esa noche se deslizaron por las calles, pasando por edificios destruidos y tropezando con los escombros. Cada sombra parecía un soldado, cada ladrido de perro les sobresaltaba. Se escuchaban a menudo llantos que provenían de las casas. El olor de la pólvora estaba en el aire. Milek deslizó su mano en la de Eva. Se le habían secado las lágrimas y caminaba decidida. Él también quería ser valiente, pero cuando miró hacia atrás, se agarró con más fuerza a su mano.

Huyeron por el bosque. *Tata* los guio entre los árboles hacia las afueras del pueblo. Las campanas de la iglesia sonaban en la oscuridad. Mamá se detuvo de repente para abrazar a Milek.

—Medianoche. Es tu cumpleaños –le susurró. Acababa de cumplir ocho años.

Anton abrió la puerta y los condujo rápidamente dentro. Los llevó arriba, al desván. Una niña, algo mayor que Munio, estaba allí, acurrucada en una esquina del cuarto.

—Es Zipora –dijo Anton. Eva le acarició el pelo. Milek se quedó mirándola. Tenía los ojos muy abiertos y asustados, y él sintió un nudo en la garganta.

Habían destrozado la familia de Eva, y luego la de Zipora. ¿Sería su familia la próxima?

Como si pudiera leer sus pensamientos, Anton dijo:

—Vamos a preparar un escondite para vosotros. Un lugar donde nadie os pueda encontrar nunca. Excavaremos y haremos una nueva casa para vosotros bajo tierra.

Noche tras noche, mientras los otros se escondían en el desván, *Tata* y Anton salían al jardín. En el sótano, donde Anton guardaba las cebollas y las patatas, cavaron un pequeño túnel

desde una esquina, hacia abajo, en dirección a la casa. Al final del túnel excavaron otra habitación, suficientemente grande para que cupieran seis personas.

—Quiero ayudar –dijo Milek.

—Yo también –exclamó Munio.

—Es demasiado peligroso –objetó mamá–. Os quedareis en el desván con nosotros.

Tata y Anton trabajaban en silencio, escuchando siempre con atención si oían alguna patrulla de soldados, o pasos de gente del pueblo deseosa de espiar para los nazis.

Anton y *Tata* cavaron con cucharas y tazas. Cuando la tierra se apilaba, se la llevaban en los bolsillos para esparcirla por el bosque, donde nadie se diera cuenta.

Al fondo de la habitación subterránea hicieron un hueco estrecho, justo debajo del dormitorio de Anton.

Anton montó una polea con una cuerda, la enganchó y la colgó en el hueco. Después cubrió el agujero. Si algún día entraba alguien no lo podría ver.

—¿Por qué a Anton le llaman tonto? –preguntó Munio–. No es tonto, es inteligente. Y, además, valiente.

Milek no dijo nada, recordando lo que antes pensaba de Anton.

Finalmente el escondite estaba listo. Anton les dio unas mantas y una lámpara de queroseno.

—Vamos a sobrevivir a esta guerra –prometió.

Munio y Milek corrieron para entrar en la nueva casa. Pero su entusiasmo pronto se desvaneció. El escondite era un agujero

húmedo. Las paredes estaban hechas de tierra sucia sostenida por unos pocos tablones de madera. El aire olía a moho.

—¡Huele a gusanos! –dijo Milek, arrugando la nariz–. ¡No me gusta!

—¿Tenemos que vivir aquí? –preguntó Munio.

Zipora no dijo nada, pero su expresión era de disgusto. Una vez que Eva, mamá y *Tata* entraron a rastras después de ellos, no había sitio para moverse. Sólo Milek era suficientemente bajito para poder ponerse de pie.

—Voy a salir –comentó Milek.

—No –dijo *Tata*–. Tenemos que quedarnos dentro.

—Pero no hay sitio. No puedo *respirar*.

Tata lo tomó en brazos.

—Sé que es difícil para ti. Es difícil para todos nosotros. Pero no podemos salir. Nos verían y se lo dirían a los nazis. Se os llevarían, hijos míos, y quizás a mamá y a mí, y a Eva y a Zipora también. Y castigarían a Anton por habernos ayudado. Tenemos que ser valientes y no dejar que nos descubran.

—¿Por qué alguien querría avisar a los nazis? –preguntó Milek–. Conocemos a todo el mundo. ¿Por qué alguien lo haría? ¿Por qué nos odian?

—Milek, algunas personas… –*Tata* empezó a hablar, pero se detuvo–. No lo sé –dijo–. Sencillamente, no lo sé.

Anton escondió la salida del túnel debajo de un montoncito de heno, para que, a pesar de todo, pudiera pasar el aire. Por la noche bajaba tres cubos desde su cuarto: uno con comida, otro con agua y otro para usarlo como retrete. Dormían en tres hileras con los pies tocándose.

Una noche, empezaron a sentir picores y a rascarse. Su escondite había sido invadido por los piojos, unos insectos que viven de chupar sangre, y estaban llenos de picadas. Munio y Milek hicieron un concurso para ver quién podía estrujar más piojos, pero no importaba cuántos contaran, ya que siempre había más.

Después de unos días, Munio preguntó:

—¿Podemos al menos jugar en el sótano?

—No –respondió *Tata*–. Los nazis están capturando judíos por todas partes. Es demasiado peligroso. Lo siento. Tenéis que quedaros aquí.

—Pero es tan *aburrido* –se quejó Milek–. ¡No hay nada que hacer!

—Podemos jugar con cordeles –afirmó Zipora–. Se quitó la cinta del pelo, ató los extremos y les enseñó cómo hacer formas con ella. Se inventaron adivinanzas que hasta a *Tata* le costaba acertar, jugaron a hacer rimas con mamá. Y cada día nombraban a las personas que faltaba de la familia de Eva y de Zipora.

—¿Podemos decir el nombre de Anton también? –preguntó Milek un día–. Ahora es como si fuera de nuestra familia y, además, está solo. ¿Qué pasará si se asusta?

—Tienes razón –observó mamá, abrazándolo.

Por la noche, casi susurrando, *Tata* les contaba historias de un mundo sin guerra en el que todos eran felices, la gente bailaba al sol y comía pasteles de miel y bollos. Mientras Milek escuchaba, con el heno de las paredes de tierra hacía dibujos que ilustraban lo que *Tata* contaba.

Fuera, la guerra seguía, con los zumbidos de los aviones y el ruido de metralla. El verano se convirtió en otoño, y luego en

invierno, y después en primavera. Bajo tierra, en el escondite de Anton, estaban ellos, entumecidos y enfermos a causa de la humedad. Las articulaciones se les hinchaban y les dolían. Cada vez resultaba más difícil soñar con una vida en la superficie y, para los chicos, más arduo acordarse de la cara de su bienhechor. Ahora Milek esperaba con ansia cuando Anton les llamaba suavemente: «Queridos, ¿cómo estáis hoy?», unas palabras que les confortaban a todos.

La comida era escasa, y cada día Anton tenía que ir más lejos para encontrarla. Les llevaba remolachas crudas, pan duro y, a veces, agua caliente con trozos de patata. Mamá le llamaba sopa, pero eso no era sopa...

A Milek le dolía la barriga.

—Tengo hambre, mamá –lloraba.

—Lo sé, pequeño –contestó mamá–. Todos estamos hambrientos. Anton hace lo que puede para encontrar comida suficiente para nosotros. A veces él se queda sin cenar para que podamos tener un poco más.

—La gente le acecha –añadió *Tata*–. Si le ven llevando comida, sospecharán. Anton está constantemente rodeado de peligros.

—Arriesga su vida cada día para que podamos sobrevivir –comentó mamá.

—Y nunca se rinde –afirmó Eva acariciando la mejilla de Milek.

Milek lloró en silencio e intentó no quejarse más.

Una noche, una voz fuerte se oyó fuera.

—¡Suchinski! Sé que escondes judíos.

Milek y Munio reconocieron la voz. Era el vecino que había amenazado a Anton meses atrás. *Tata* rápidamente sopló para que se apagara la lámpara. Les dio trozos de tela para que se los pusieran en la boca para que no hicieran ruido. Nadie se atrevía a moverse lo más mínimo.

—Los nazis pagan una buena suma de dinero por lo judíos. Quinientos zlotis por un adulto y mil por un niño. Entrégalos y nos repartiremos el dinero.

—No escondo a nadie –respondió Anton.

—¡Embustero! –vociferó el hombre–. Apuesto a que son los mocosos que estaban en tu patio aquel día. ¡Te lo advierto! Se lo contaré a los soldados y los perros pronto olisquearán a tus queridos amigos. ¡Y yo me quedaré el dinero para mi solo!

Aquella noche, Anton echó estiércol mezclado con hierbas y pimienta por el suelo y por los peldaños que conducían al sótano.

Cuando empezaba a amanecer, el vecino llegó acompañado de soldados.

—¿Dónde están los judíos? –gritó el capitán–. ¡Sabemos que los estás escondiendo! Dinos dónde, o te arrepentirás.

—Aquí no hay nadie más que yo –aseguró Anton.

—¡Habla! –ordenó el capitán, apoyando el fusil en el pecho de Anton.

Anton pensó en las seis personas bajo tierra, cada una de ellas una vida valiosa. Miró al soldado a los ojos.

—No hay nadie más que yo.

—¡Miente! –exclamó el vecino.

—¡Buscadlos! –ordenó el capitán. Los soldados registraron la casa. Vaciaron los armarios y apartaron los muebles, revolvieron todo el desván. Después bajaron al sótano.

Milek cerró los ojos con fuerza, convencido de que estaban a punto de ser capturados. Los perros olfateaban y olfateaban, pero el estiércol y las hierbas que Anton había esparcido les en-

gañaban el olfato. Y el heno que Anton había colocado encima de la entrada del túnel lo mantenía escondido. Los soldados no encontraron nada.

—Te sacaremos la verdad —dijo el capitán—. ¡Si no, morirás!

Los soldados se llevaron a Anton.

Anton estuvo ausente el resto del día, y el siguiente, y el siguiente.

—¿Dónde está Anton, mamá? –preguntaba Milek–. ¿Está a salvo? –Mamá no tenía respuestas y Milek lloraba, inquieto por su amigo.

Después de cinco días, un grupo de soldados acamparon en el jardín de Anton. Dormían en el sótano. La luz de una linterna se filtraba a través del heno mientras bromeaban y discutían. Milek olió algo –¿era chocolate?– y se le hizo la boca agua. ¿Cuánto tiempo hacía que no probaba nada parecido? Oían el sonido metálico de las armas cuando las limpiaban y las recargaban.

—¡Ah! –dijo un soldado–. Se me ha caído una bala.

Una pequeña pieza de metal cayó en el túnel y rodó hacia abajo, hasta los pies de Munio. Éste se quedó petrificado de miedo.

—¿Adónde ha ido a parar? Tiene que estar en alguna parte. –Se oía el roce de los pies de los soldados con el heno mientras buscaban.

El corazón de Milek latía con fuerza. Podían descubrirlos en cualquier momento. Se los llevarían a todos. Mordió el trapo con fuerza para no chillar.

Se oyó un grito fuera y los soldados se marcharon deprisa. Hubo más gritos más lejos, pero Milek no entendía las palabras. Se oyeron tiros. Después llegó el silencio. ¿Qué estaba pasando? ¿Había regresado Anton? ¿Le habían disparado los soldados?

De repente, se oyeron pasos. Alguien había entrado en el sótano. Los pasos eran cada vez más fuertes. Retiraron el heno. Una luz intensa deslumbró el escondite. Milek escondió la cara en el regazo de mamá. ¡Los nazis los habían encontrado!

Pero, tras la luz, oyeron una voz dulce y familiar.

—¡Queridos! ¡Oh, queridos, salid! ¡Ya podéis salir, ya estáis a salvo! ¡Los nazis se han marchado!

Munio se quitó el trapo de la boca.

—¡Anton! –gritó.

Se arrastró hacia la salida del túnel, a través del sótano, hacia el exterior. Le siguieron Milek y los otros. Estaban tan débiles, después de estar encogidos bajo tierra tanto tiempo, que Anton les tuvo que ayudar a cada uno de ellos a salir. Ya de pie, en el exterior, parpadeaban a causa de la luz de un día claro de verano. Sintieron que les calentaba la cara. Respiraron el aire limpio. ¿Eran realmente libres, después de esos largos meses?

Con lágrimas en los ojos, mamá y *Tata* dijeron a Anton:

—¡Nunca podremos darte las gracias por lo que has hecho!

—Pensamos que te habían matado –dijo Eva. Anton les abrazó a todos.

Milek lo miró con admiración.

—¡Nos has salvado!

Anton sonrió.

—Queridos, ¿qué otra cosa hubiera podido hacer? Mi corazón sabía que era lo que debía hacer. La vida tiene un valor inmenso, cada una de ellas. –Tomó la mano de Milek y la estrechó suavemente entre las suyas.

¿Qué pasó después?

En julio de 1944, los nazis abandonaron Zborów, el pueblo de Milek y Munio (entonces pertenecía a Polonia, ahora a Ucrania). Los seis supervivientes permanecieron con Anton, recuperándose lentamente del año que estuvieron escondidos hasta que terminó la guerra en 1945.

Milek y Munio se trasladaron a Estados Unidos en diciembre de 1949. Los chicos se cambiaron los nombres: Milek se llamó Shelley y Munio se puso de nombre Michael. Los dos hermanos viven todavía en Estados Unidos con sus esposas, hijos y nietos.

Ésta es la entrada que conducía al escondite.

Al principio de la guerra, Eva se había comprometido con un hombre que había ido a buscar trabajo a Uruguay, en América del Sur. Acabada la guerra, su prometido la buscó a través de la Cruz Roja y la comunidad judía. Se reunieron en Uruguay. Eva se casó y tuvo dos hijos, Emmanuel y Baruch. La familia más tarde se trasladó a Israel. Hoy, Emmanuel y su esposa, reparten su tiempo entre el trabajo en Canadá y la familia en Israel. Baruch vive con su familia en Israel, donde también trabaja.

La familia Zeiger después de la guerra.

Eva Adler murió en 2005.

Zipora Stock se marchó a Israel, donde se casó. Ella, sus hijos y sus nietos viven todavía allí.

Los Zeiger intentaron convencer a Anton para que se fuera con ellos a Estados Unidos, pero él no quiso marcharse. Les dijo que «quería morir donde nació». Durante años, la señora Zeiger le mandaba paquetes de comida enlatada y ropa. Como Anton no sabía leer ni escribir, dictaba sus cartas a un vecino. En cada carta dibujaba una flor, para que los Zeiger supieran que la carta era suya.

Durante la década de 1960, las cartas de Anton dejaron de llegar. Los Zeiger hicieron todo lo posible para localizarlo, pero era difícil tener noticias de Zborów. Los miembros de la Cruz Roja lo buscaron, pero no pudieron localizarlo. Su casa estaba vacía y nadie sabía qué le había ocurrido. Con mucha tristeza, los Zeiger acabaron pensando que había muerto.

El señor Zeiger (*Tata*) falleció en 1971.

En 1988, Shelley (Milek) Zeiger conoció a un hombre de la región de Zborów que accedió a intentar buscar a Anton por última vez. En este caso la búsqueda fue un éxito. Los Zeiger descubrieron que Anton vivía sumido en la pobreza. Le construyeron una casa nueva y se las arreglaron para que tuviera a alguien que pudiera cuidar de él durante el resto de su vida.

Al año siguiente, Shelley, Michael y su madre viajaron a Zborów y se reunieron con el hombre que les había salvado la vida. Unas mil personas del pueblo festejaron el acontecimiento con los Zeiger y Anton. «El tonto del pueblo» se había convertido en el héroe del pueblo.

Finalmente, en 1988, después de años de búsqueda, un conocido de Shelley encontró a Anton. Todos creían que había fallecido.

Shelley (Milek), mamá y Michel (Munio), en 1988, cuando los Zeiger viajaron a su pueblo para ver de nuevo a Anton.

Los habitantes de Zborów, con mamá y Anton, durante la fiesta que se celebró en su honor. Anton se había convertido en el héroe del pueblo.

*Después de la reunión en
Zborów, en 1989, Anton
y mamá (la señora Zeiger)
se volvieron a encontrar
cuando Anton los visitó
en Estados Unidos.*

Ahora hay un monumento erigido por la familia Zeiger a los judíos asesinados de Zborów.

En 1992 se volvieron a encontrar en la celebración con Michael Zeiger, Eva Adler, Zipora Stock (ahora Shindelheim) y sus familias. Anton Suchinski fue reconocido como Justo entre las Naciones, Un Gentil Justo. En El Yad Vashem, Museo Memorial del Holocausto en Jerusalén. Su nombre fue añadido a la lista del Muro de Honor del Jardín de los Justos.

Eva, Anton y Zipora, con Michael detrás de ellos, en la ceremonia en el Yad Vashem en 1992.

Michael, Eva, Zipora y sus familias alrededor de Anton. Tras ellos, el Muro de Honor en el Jardín de los Justos.

Zipora y Eva sonríen mientras Anton enseña el Certificado de Honor.

La señora Zeiger (mamá) murió en 1997.

Anton Suchinsky falleció en 2001 a la edad de noventa y seis años.

La medalla que Anton recibió en el Yad Vashem tiene grabado un dicho judío que reza lo siguiente: «Quien salva una vida, salva al mundo entero».

El nombre de Anton fue incluido en el Muro de Honor en el Jardín de los Justos.

Para leer más:
Zeiger, Shelley y Maryann McLoughlin, *The Weel of Life: A Memoir*, Margote, Comteq, 2012.
www1.yadvashem.org/yv/en/righteous/stories/sukhinski.asp